PATTERNS IN STONE

IF WE COULD turn back the clock 300 million years and explore the ancient seas of the Paleozoic Era, we would discover a bewildering array of animal life. There would be terrifying predators, like *Dinichthys*, a gigantic fish with large, scissor-like jaws in a heavily armored skull; and ammonoids, molluscs resembling squids but housed in elegant, chambered shells. On the sea floor, we would find some familiar-looking animals, like snails, sponges, corals, and sea-stars. Much less familiar and anchored to the bottom would be hard-shelled creatures with fleshy stalks called brachiopods and vast gardens of sea-lilies (crinoids) with their feathery arms swaying in the current.

We would also find trilobites, scurrying along the bottom searching for food in the primordial ooze. These strange anthropods with segmented, external skeletons and jointed legs, are extinct today but were enormously successful in the Paleozoic Era. They varied in size, from barely visible to nearly a meter (39 inches) long, and in shape, from conservative and streamlined to bizarrely ornamented with knobs and long spines. Some trilobites were blind; but most of them, like their insect relatives, possessed compound eyes, some with as many as 15,000 lenses.

Although time travel is beyond our reach, fossils offer us a window on the long prehistory of life on Earth. Life began in the sea. Single-celled organisms, alone, reigned supreme for about 80% of the time during which life is known to have existed on this planet. As unassuming as they may seem, these early organisms drastically altered the Earth's environment and allowed life to emerge as we know it today.

The first traces of life, bacteria, are preserved as microscopic fossils in marine sediments laid down 3.6 billion years ago. Sometime in the first billion years, blue-green algae appeared. These organisms had the ability to employ the energy of sunlight to transform carbon dioxide and water into food and release oxygen as a waste product. This miraculous process of photosynthesis, still used by green plants today, enriched the atmosphere and seas with oxygen and thus permitted the development of new life-forms.

The first multicelled organisms entered the fossil record a mere 700 million years ago. Eventually, some developed sexual reproduction, a major breakthrough that allowed greater variability in genetic makeup and accelerated evolutionary change. Throughout the Precambrian Era, life was limited to only a few types of plants and animals. But something wonderful was about to happen.

ANCIENT SEA CREATURES

A Wild Horizons® Postcard Book

Photography by Thomas A. Wiewandt
Introduction by Peter L. Larson

WILD HORIZONS PUBLISHING, INC.
TUCSON, ARIZONA

Series Editor: Thomas Wiewandt
Design: Paul Mirocha
Production: Ejyo Katagiri
Copy Editing: Sally Antrobus-Wilkinson,
    Jeffrey H. Lockridge
Author's Bio: Deborah Mitchell
Scientific Consultants:
    Peter L. Larson & Robert A. Farrar,
      Black Hills Institute of Geological Research
    Dr. Stanley S. Beus, Dept. of Geology,
      Northern Arizona University
Fossil Specimens: Courtesy of the individuals and
      organizations credited in each caption

Publisher's Cataloging-in-Publication Data:

Wiewandt, Thomas A.
ANCIENT SEA CREATURES
    A Wild Horizons Postcard Book

ISBN 1-879728-02-8

1. Fossils, marine—pictorial works.
2. Extinct animals—pictorial works.
3. Evolution (biology)—pictorial works.
4. Paleontology—pictorial works.   I. Title
    QH or QE or TR 1992

First Edition, 1992, Printed in Hong Kong 7 6 5 4 3 2

Wild Horizons Publishing
Post Office Box 5118-02
Tucson, Arizona 85703-0118
U.S.A.

For more Wild Horizons books,
try your local booksellers,
call toll free: 1-800-795-1513,
or send a FAX: 602-798-1514.

Suddenly, about 570 million years ago, a wild proliferation of new life-forms arose. It was as though a gate had swung open and the seas were flooded with a vast menagerie of strange, new creatures. Because this event can be so clearly seen in fossil-bearing rocks throughout the world, geologists have used it to define the boundary between the first two great eras in the history of life, the Precambrian and the Paleozoic.

Many of these new animals that marked the onset of the Paleozoic Era (beginning with the Cambrian Period) had hard external skeletons, which strongly favored their preservation as fossils. But scientists since the days of Charles Darwin have remained puzzled by the speed with which life achieved such complexity and diversity at this time. Just 70 million years after the Paleozoic had begun, virtually all modern groups (phyla) of animals were present on Earth. And by the end of the next 90 million years—through the middle of the Paleozoic—insects, scorpions, and plants had colonized the land.

The fossil record also reveals periodic mass extinctions of immense proportions. The boundary between the Paleozoic and the Mesozoic eras, about 240 million years ago, was marked by the extinction of more than 90% of the marine species. And the Mesozoic Era was brought to a close by another mass extinction, which included the dinosaurs. Perhaps 70% of all Mesozoic plants and animals, on land and in the seas, became extinct before the beginning of the current chapter of life, the Cenozoic Era.

But what of our ancestors? Early Paleozoic rocks document the arrival of a worm-like creature called *Pikaia*. Paleontologists believe that this small organism gave rise to all fishes and to one very special group, the lobed-finned fishes, with stout, fleshy fins, some of which left the Paleozoic seas to live on land. From these fishes came a series of land vertebrates, eventually including humans. Thus, we, too, have our beginnings in those ancient seas.

The symmetry of life, depicted in the intricacy of fossil specimens, has absorbed human thought for thousands of years. Why some forms flourished and others vanished remains a mystery. We continue to ponder the significance of these "patterns in stone" and cherish them as objects of beauty and inspiration.

— PETER L. LARSON

Peter Larson is President of the Black Hills Institute of Geological Research, Hill City, South Dakota. Since childhood, he has been digging, preparing, and studying fossils—everything from ammonites and dinosaurs to saber-tooth cats. He and his Black Hills colleagues have supplied specimens to many of the world's great museums.

GEOLOGIC TIME CLOCK

Depicting the 4.5 Billion Years of the Earth's Existence in Twelve Hours

. . . .

PRECAMBRIAN ERA
Age of Micro-organisms
4.5 billion - 570 million years ago

PALEOZOIC ERA
Age of Trilobites, Crinoids & Brachiopods
570 - 240 million years ago

MESOZOIC ERA
Age of Ammonites and Dinosaurs
240 - 65 million years ago

CENOZOIC ERA
Age of Insects, Birds, & Mammals
65 million years ago to present

Immediate ancestors of humankind emerge during the last 45 seconds of our 12-hour time clock.

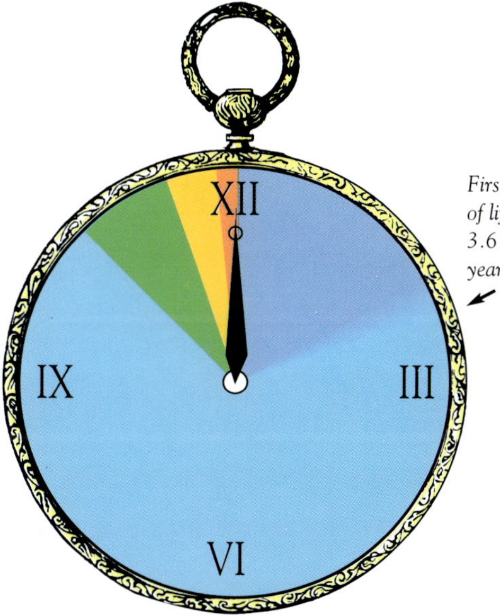

First traces of life occur 3.6 billion years ago.

FOSSILS REVEAL THAT single-celled organisms ruled alone and supreme for nearly 80% of the history of life on our planet. An astonishing array of complex life-forms made a dramatic entrance 570 million years ago, in Paleozoic seas, but most perished by the end of that era. The next great era, the Mesozoic—the age of ammonites and dinosaurs—ended 65 million years ago with another mass extinction, ushering in the most recent era, the Cenozoic. Just 4–5 million years before present, human-like creatures emerged.

From *ANCIENT SEA CREATURES*, a Wild Horizons Postcard Book, P.O. Box 5118-02, Tucson, Arizona 85703-0118 USA
© 1991 by Wild Horizons Publishing; Illustration by Paul Mirocha.

STROMATOLITES ARE PROBABLY the oldest fossils that can be seen with the naked eye. These columns of blue-green algae interlayered with silt were deposited more than 2 billion years ago, in warm Precambrian seas. Replacement of algae by quartz and iron oxide gives them a red color, as in this cross-section from Minnesota. Living stromatolites occur in only a few places worldwide. (102)

From *ANCIENT SEA CREATURES*, a Wild Horizons Postcard Book, P.O. Box 5118-02, Tucson, Arizona 85703-0118 USA
Photo © 1991 by Thomas Wiewandt. Specimen courtesy of Dr. Klaus Westphal, Museum of Geology, University of Wisconsin, Madison Campus.

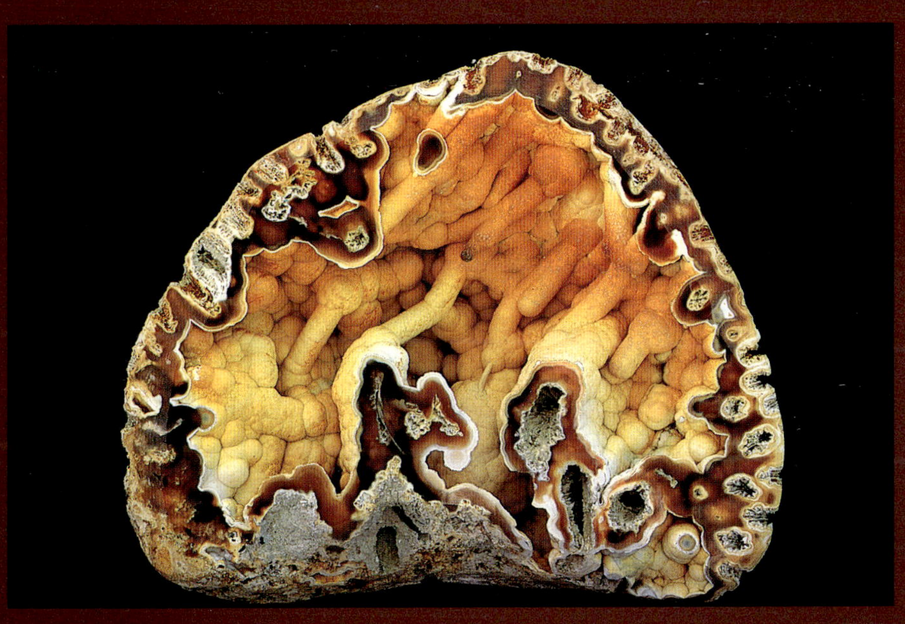

CORALS THRIVED in seas of the Paleozoic but declined sharply by the end of that era. Not long after, early in the Mesozoic, new forms arose and began building extensive reefs. This agatized coral head, cut in half, is of Cenozoic (Miocene) age, from Tampa Bay, Florida. (202)

IN SOME PALEOZOIC marine sediments, lampshells (brachiopods) are the most common fossils; about 30,000 species have been found. We know from the few modern survivors that brachiopods were stationary filter-feeders. These shells of *Mucrospirifer mucronatus*, mineralized with pyrite, are from the Middle Devonian Silica Shale of Sylvania, Ohio. (302)

From *ANCIENT SEA CREATURES*, a Wild Horizons Postcard Book, P.O. Box 5118-02, Tucson, Arizona 85703-0118 USA
Photo ©1991 by Thomas Wiewandt. Specimen courtesy of Jon Kramer, Potomac Museum Group, Golden Valley, Minnesota.

ALTHOUGH STALKED and typically anchored to the ocean floor, sea-lilies (crinoids) are not plants. They are relatives of sea-stars and sea-urchins (echinoderms) and use their spreading, flower-like arms to comb the water for food particles. These *Cyathocrinites gorbyi* (left) and *Macrocrinus mundulus* (right) are of late Paleozoic age, from Montgomery Co., Indiana. (402)

From *Ancient Sea Creatures*, a Wild Horizons Postcard Book, P.O. Box 5118-02, Tucson, Arizona 85703-0118 USA
Photo © 1986 by Thomas Wiewandt. Specimen courtesy of Robert Howell, Jr., Geoscience Enterprises, Roachdale, Indiana.

S OME SEA-LILIES (crinoids) floated free and must have looked like armored jellyfish. A few descendants from these drifters of the open ocean and their anchored counterparts are living today. This animated swarm of *Uintacrinus socialis* was uncovered from the late Mesozoic (Cretaceous) Mancos Shale in Colorado. (502)

From ANCIENT SEA CREATURES, a Wild Horizons Postcard Book, P.O. Box 5118-02, Tucson, Arizona 85703-0118 USA
Photo © 1991 by Thomas Wiewandt. Specimen courtesy of William Hawes, Jr., Grand Junction, Colorado.

AMONG THE MOST ADVANCED and diverse animal groups early in the Paleozoic were trilobites, bottom dwellers named for their three-lobed bodies. Distantly related to insects, crabs, and spiders, they are dubbed "bugs" by fossil collectors. This plate of Ordovician shale with six *Pseudogygites latimarginatus* is from Bowmanville, Ontario, Canada. (602)

From *ANCIENT SEA CREATURES*, a Wild Horizons Postcard Book, P.O. Box 5118-02, Tucson, Arizona 85703-0118 USA
Photo © 1991 by Thomas Wiewandt. Specimen courtesy of the Black Hills Institute of Geological Research, Hill City, South Dakota.

TRILOBITES WERE AMONG the first animals to see the world around them. Although some were blind, many had well-developed compound eyes, with up to 15,000 separate lenses. This *Huntonia oklahomae*, with crescent-shaped eyes, is from the Paleozoic (Devonian) limestone of Coal Co., Oklahoma. (702)

From *Ancient Sea Creatures*, a Wild Horizons Postcard Book, P.O. Box 5118-02, Tucson, Arizona 85703-0118 USA
Photo © 1991 by Thomas Wiewandt. Specimen courtesy of Robert Carroll, Ann Arbor, Michigan.

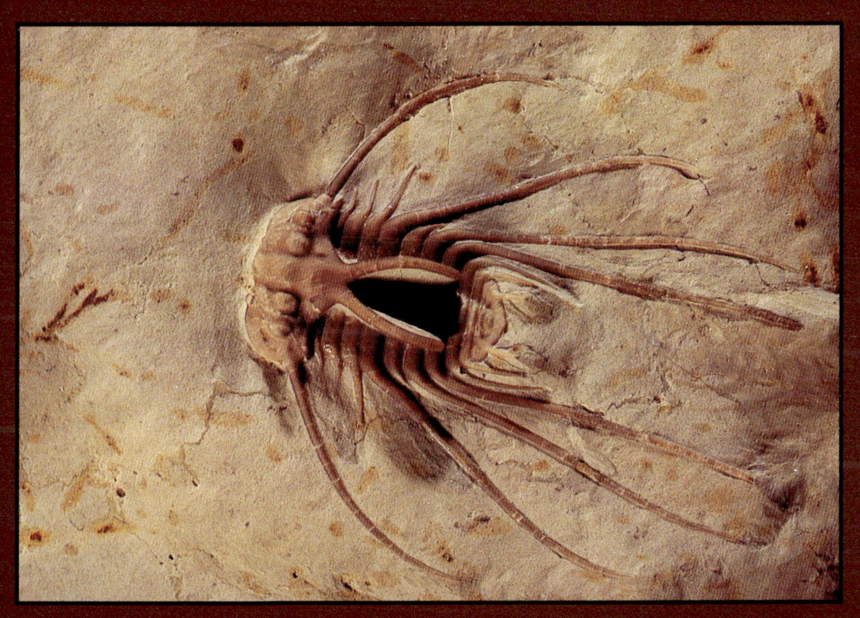

THIS ORNATE TRILOBITE, *Dicranurus elegantus*, collected in the Arbuckle Mountains of Oklahoma, roamed the ocean floor during the middle of the Paleozoic (Devonian), more than 350 million years ago. Because of its long, delicate spines, complete specimens are extremely rare and tedious to prepare. (802)

From ANCIENT SEA CREATURES, a Wild Horizons Postcard Book, P.O. Box 5118-02, Tucson, Arizona 85703-0118 USA
Photo © 1986 by Thomas Wiewandt. Specimen courtesy of Geological Enterprises, Ardmore, Oklahoma.

TRILOBITES PERISHED before the Age of Dinosaurs. Perhaps they fell prey to jawed fishes and other large predators, which increased in diversity as trilobites declined. Masses of these *Homotelus bromidensis*, found in Oklahoma, died together—did they meet a catastrophic end while breeding, suffer from a population explosion, or get trapped in a shrinking puddle? (902)

From *ANCIENT SEA CREATURES*, a Wild Horizons Postcard Book, P.O. Box 5118-02, Tucson, Arizona 85703-0118 USA
Photo © 1991 by Thomas Wiewandt. Specimen courtesy of George Lee, Costa Mesa, California;
from Geological Enterprises' quarry, Ardmore, Oklahoma.

SEA-STARS (STARFISH) have changed little in appearance during the past 400 million years and continue to be well represented in modern marine communities. This unidentified species of *Furcaster* was collected from the Hunsrück Shale beds of mid-Paleozoic (Early Devonian) age near Bundenbach, Germany. (1002)

From *Ancient Sea Creatures*, a Wild Horizons Postcard Book, P.O. Box 5118-02, Tucson, Arizona 85703-0118 USA
Photo © 1991 by Thomas Wiewandt. Specimen courtesy of Flavio Bacchia, Coelodus Arte Natura, Trieste, Italy

ANIMALS COMMONLY KNOWN as snails (gastropod molluscs) flourished in ancient seas. Once filled with mud, the shell of this *Turritella mortoni* dissolved away, leaving only a stone cast (steinkern) of its corkscrew-shaped interior. This specimen is from early Cenozoic deposits found near Indian Head, Maryland. (1102)

From *Ancient Sea Creatures*, a Wild Horizons Postcard Book, P.O. Box 5118-02, Tucson, Arizona 85703-0118 USA
Photo © 1991 by Thomas Wiewandt. Specimen courtesy of the Potomac Museum Group, Golden Valley, Minnesota.

NAUTILOIDS AND AMMONITES were cephalopods, advanced molluscs closely related to octopuses and squids. Many had coiled shells, while others had long, tapering, conical shells, like these orthocerids from mid-Paleozoic limestone in Morocco. All ammonites are extinct; but in the seas of the Indo-Pacific, nautiloids have a lone survivor, the pearly nautilus. (1202)

From *ANCIENT SEA CREATURES*, a Wild Horizons Postcard Book, P.O. Box 5118-02, Tucson, Arizona 85703-0118 USA
Photo © 1991 by Thomas Wiewandt. Specimen courtesy of Brian Eberhardie, Moussa Direct Ltd., Cambridge, England.

PRIZED BY FOSSIL COLLECTORS are ammonites found with their outer shell layers intact, like this *Sphenodiscus lenticularis* from South Dakota. Its iridescent shell refracts light, glowing bright red and green. Among the last ammonites to survive, this species vanished with the dinosaurs about 65 million years ago, at the close of the Mesozoic Era. (1302)

From ANCIENT SEA CREATURES, a Wild Horizons Postcard Book, P.O. Box 5118-02, Tucson, Arizona 85703-0118 USA
Photo © 1986 by Thomas Wiewandt. Specimen courtesy of the Black Hills Institute of Geological Research, Hill City, South Dakota.

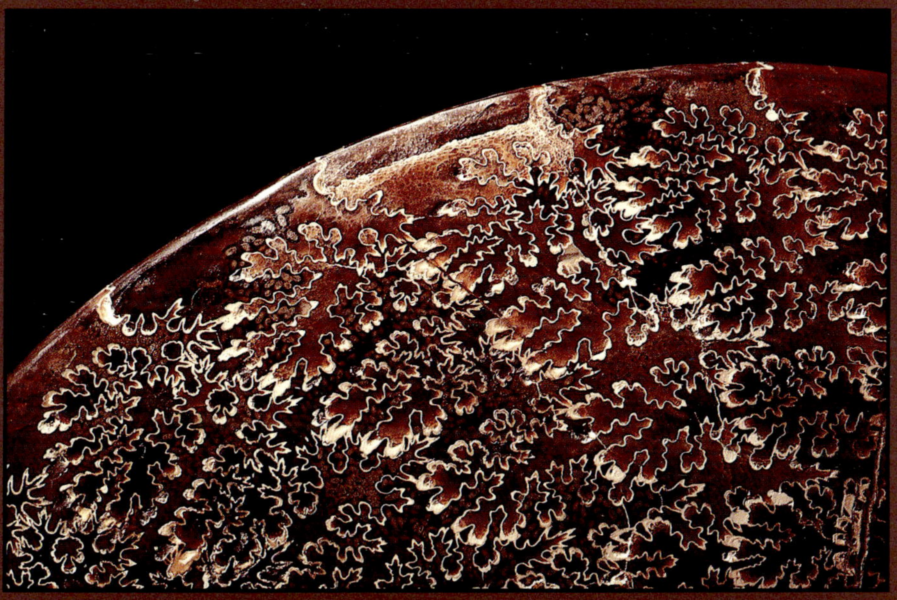

AMMONITE FOSSILS often lost their outer, mother-of-pearl shell layer, exposing an intricate pattern of suture lines formed as the animal grew. As the creature enlarged its shell, it partitioned off new chambers within. This close-up shows a maze of sutures on a *Placenticeras meeki* from late Mesozoic sediments in Meade Co., South Dakota. (1402)

From *ANCIENT SEA CREATURES*, a Wild Horizons Postcard Book, P.O. Box 5118-02, Tucson, Arizona 85703-0118 USA
Photo © 1986 by Thomas Wiewandt. Specimen courtesy of Japheth Boyce, R.J.B. Rock Shop, Rapid City, South Dakota.

Ammonite shells were often spiraled and ridged like a ram's horn. They derive their name from the horn of Ammon, supreme deity of ancient Egypt. This *Pleuroceras spinatum*, from Mesozoic (Jurassic) deposits found near Unterstürmig, Germany, has been partially mineralized with pyrite. (1502)

From *Ancient Sea Creatures*, a Wild Horizons Postcard Book, P.O. Box 5118, Tucson, Arizona 85703
Photo© 1991 by Thomas Wiewandt. Specimen courtesy of Dietmar Seeh, Galerie Fossil, Tuttlingen, Germany.

THIS CROSS-SECTION of a *Ludwigia murchisoni* shell, from Dorset, England, shows its inner whorls and once-hollow chambers, now filled with crystalline calcite. Like the pearly nautilus, which survives today, ammonites were able to control their buoyancy in submarine fashion, by adjusting the amount of liquid in each chamber. (1602)

SEA-SCORPIONS (EURYPTERIDS) were the largest of the Paleozoic free-swimming creatures with external skeletons—some grew to 3 meters (10 feet) in length and had huge pincers. Eurypterids appear to have been closely related to modern scorpions and spiders. This individual, *Eurypterus remipes*, lived more than 400 million years ago and was discovered in Herkimer Co., New York. (1702)

From ANCIENT SEA CREATURES, a Wild Horizons Postcard Book, P.O. Box 5118-02, Tucson, Arizona 85703-0118 USA
Photo © 1986 by Thomas Wiewandt. Specimen courtesy of the Black Hills Institute of Geological Research, Hill City, South Dakota.

FISHES FIRST APPEARED during the middle of the Paleozoic Era. With the development of early bony forms came some with cartilaginous skeletons—sharks, rays, and skates—which are essentially unchanged today. This skate from Lebanon, *Cyclobatis oligodactylus*, inhabited the sea floor late in the Mesozoic Era. (1802)

From *ANCIENT SEA CREATURES*, a Wild Horizons Postcard Book, P.O. Box 5118-02, Tucson, Arizona 85703-0118 USA
Photo © 1991 by Thomas Wiewandt. Specimen courtesy of the Black Hills Institute of Geological Research, Hill City, South Dakota.

Ancestors of fish that dominate fresh and salt water environments today evolved during the last third of the Paleozoic Era. This ancient fish, *Discoserra pectinodon*, swam in lagoons 320 million years ago; from the Heath Shale formation of Fergus Co., Montana. (1902)

From *Ancient Sea Creatures*, a Wild Horizons Postcard Book, P.O. Box 5118-02, Tucson, Arizona 85703-0118 USA
Photo © 1991 by Thomas Wiewandt. Specimen courtesy of William Hawes, Jr., Grand Junction, Colorado.

THIS ELEGANTLY PRESERVED fish, *Mene rhombea*, graced tropical waters of an ocean bay approximately 50 million years ago. It was found in the early Cenozoic (Eocene) fossil beds of Monte Bolca, Italy. (2002)

From ANCIENT SEA CREATURES, a Wild Horizons Postcard Book, P.O. Box 5118-02, Tucson, Arizona 85703-0118 USA
Photo © 1991 by Thomas Wiewandt. Specimen courtesy of Andreas Guhr, Mineralien Zentrum, Hamburg, Germany.

ABOUT THE PHOTOGRAPHER

AS A CHILD in northern New Mexico, Tom Wiewandt formed a deep and lasting spiritual connection with the natural world—from the lashing power of a summer storm to such small dramas as a fence lizard doing push-ups on a log.

Armed with a mastery of photography and a Ph.D. in ecology, Tom wants to pique our curiosity and sense of wonder. He blends art and science in a tapestry of images that stir our imaginations and make us long to see more. ANCIENT SEA CREATURES developed from his growing interest in fossils and his decision to live in Tucson, Arizona, host to the largest gem, mineral, and fossil show in the world.

Tom's camera has brought the mysteries of things wild to the pages of *Audubon*, *Smithsonian*, *Omni*, and *National Wildlife*. His first trade book, HIDDEN LIFE OF THE DESERT (Crown/Random House, 1990), made the John Burroughs List of Outstanding Nature Books for Young Readers. Because life doesn't stand still, Tom has captured the dynamics of animal behavior and the world around us in his films for the BBC and the National Geographic Society. He also invites others to experience the lure of field photography on his Wild Horizons Photographic Safaris, where participants learn to interact with nature from both sides of the camera.

DESERT RAIN/DESERT BLOOM
(ISBN 1-879728-00-1) 1992

Rain brings unsurpassed beauty to deserts of the American Southwest—celebrated in this book through magnificent pictures of cloudbursts, lightning, rainbows, and wildflower displays seen once in a decade. Thomas Wiewandt's photographs and John Alcock's lyrical introduction contrast the moods, cycles, and life-giving significance of rain in arid lands.

CACTUS FLOWERS
(ISBN 1-879728-01-X) 1992

Cactus plants, native to the Americas from Canada to Argentina, are prized for their unusual and extravagantly beautiful flowers. Thomas Wiewandt has captured their fleeting splendor on film, in a fiesta of colorful close-ups—from the lime green blossoms of a teddybear cholla to day-glow pinks of an orchid cactus. The introductory text by Mark Dimmitt gives a lucid overview of cactus family.

ALSO AVAILABLE

. . . .

DECOR PRINTS

All of the pictures in this book are available as limited edition, signed and numbered, custom-made photographic enlargements of archival quality. If you would like information and prices, please list the image number given in parentheses at the end of the caption on the back of each photo you've selected:

___________, ___________, ___________, ___________

WILD HORIZONS® PHOTOGRAPHIC SAFARIS

At Wild Horizons, we take you beyond fine photography and the excitement of science—to our source of inspiration, to the best classroom of all, the world's wild places.

Since 1985, we have been offering first-class learning vacations to spectacular destinations in the American West and abroad. These all-inclusive, 1- to 3-week tours are led by professionals and feature natural history interpretation with hands-on photographic instruction. Our groups are kept small and our pace leisurely, with ample opportunity to refresh your senses and expand your creativity in an unhurried atmosphere of fun and relaxation.

If you would like to be added to our mailing list, please indicate your preferences below.

❏ 1- to 2-week tours of natural wonders in the American West

❏ 2- to 3-week tours abroad, to E. Africa, the Galápagos, the Amazon Basin, or similarly wild places that can be experienced in comfort

We also design tours for private groups of 4-8, arranged at least 9 months in advance (sorry, no tours to Europe).

WE VALUE YOUR OPINION. Please tell us what you think about our postcard books:

____

____

_________________ May we quote you? ________ If yes, please include your phone number (________)_______________

Please print your name and address in the upper left corner on the reverse side of this reply card.

Wild Horizons
P. O. Box 5118-02
Tucson, Arizona 85703-0118
U. S. A.